Risaku Suzuki

Snow Letter

Nazraeli Press

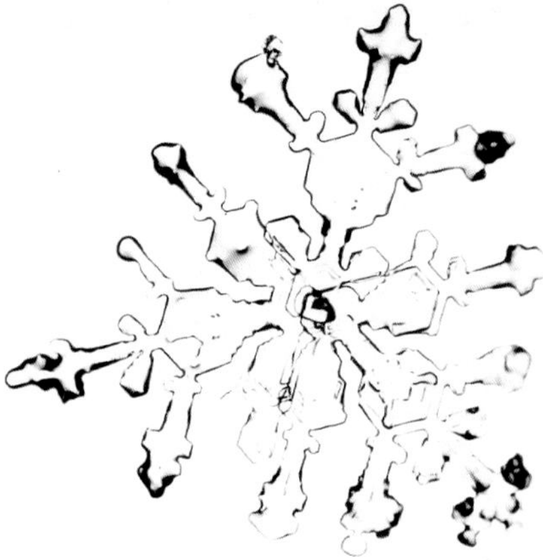

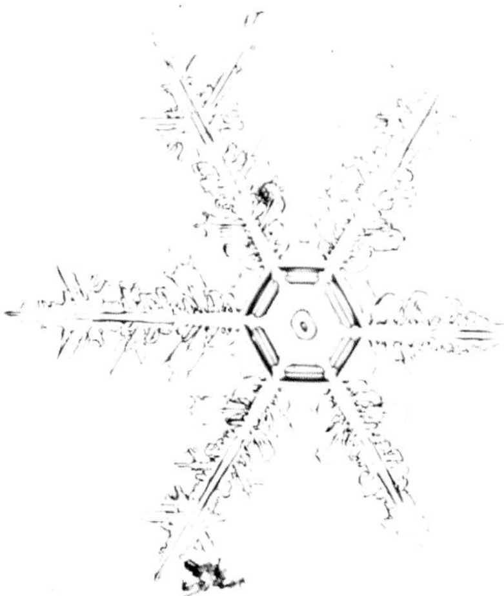

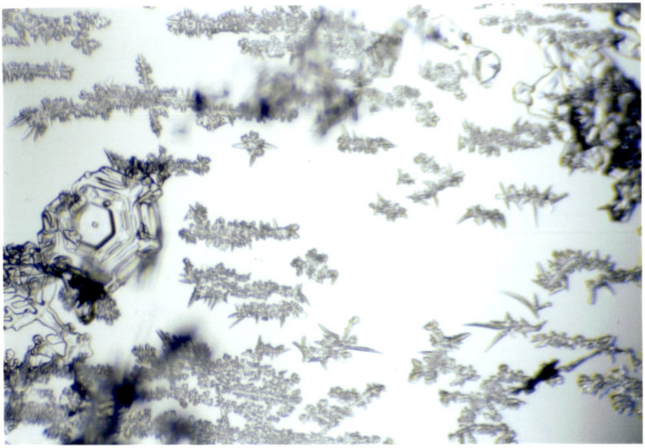

Snow Letter
A book of 11 reproductions, and one original photograph
by Risaku Suzuki.

N° 275 / 500 (Print signed on verso)

Nazraeli Press One Picture Book N° 80
Printed in an edition of 500 copies

Order Address: Nazraeli Press
2871 NE Alameda Street
Portland, Oregon 97212, USA

Printed in China for Nazraeli Press
ISBN 978-1-59005-380-5